THE AMOROUS

CHRIS WALLACE-CRABBE

The Amorous Cannibal

Oxford   New York
OXFORD UNIVERSITY PRESS
1985

Oxford University Press, Walton Street, Oxford OX2 6DP
London New York Toronto
Delhi Bombay Calcutta Madras Karachi
Kuala Lumpur Singapore Hong Kong Tokyo
Nairobi Dar es Salaam Cape Town
Melbourne Auckland
and associated companies in
Beirut Berlin Ibadan Mexico City Nicosia

Oxford is a trade mark of Oxford University Press

© *Chris Wallace-Crabbe 1985*
First published 1985 by Oxford University Press

All rights reserved. No part of this publication may be reproduced,
stored in a retrieval system, or transmitted, in any form or by any means,
electronic, mechanical, photocopying, recording, or otherwise, without
the prior permission of Oxford University Press

This book is sold subject to the condition that it shall not, by way
of trade or otherwise, be lent, re-sold, hired out or otherwise circulated
without the publisher's prior consent in any form of binding or cover
other than that in which it is published and without a similar condition
including this condition being imposed on the subsequent purchaser

British Library Cataloguing in Publication Data
Wallace-Crabbe, Chris
The amorous cannibal.—(Contemporary poetry)
I. Title II. Series
821   PR9619.9.W28
ISBN 0-19-211968-0

Library of Congress Cataloging in Publication Data
Wallace-Crabbe, Chris.
The amorous cannibal.
I. Title.
PR9619.3.W28A83 1985   823   84-29481
ISBN 0-19-211968-0 (pbk.)

Set by Rowland Phototypesetting Ltd,
Printed in Great Britain by
J. W. Arrowsmith Ltd, Bristol

Contents

The Amorous Cannibal   1
Mind   2
The Figure in the Carpet   3
Recollection   5
The Landlord of Himself   6
Forgetting   7
Abhorring a Vacuum   8
Smart as Paint, or Less   9
The Shadow Minister   10
Nub   11
Hoffmann Drunk   12
Exit the Players   13
Spirit and Act, or the Last Metro   14
Squibs in the Nick of Time   15
Redemptive Heights   16
Panoptics   17
The Good Spirit Bounces Back   18
The Bits and Pieces   19
Puck Is Not Sure about Apollo   31
The Fall of the West   32
Eating the Future (I)   33
Eating the Future (II)   34
Words   35
That Radical Politics Is Impossible   36
Sand in My Shoes   37
Practitioners of Silence   38
We'll Build a Stairway to Paradise   39
The Home Conveyancing Kit   40
Patient Portent   41
River   42
Kia Ora   43
Shadows   45
Aunt   46
Under the Roof of the World   48
Gaspard de la Nuit   49
Still   50
The Mower against Ocean   51
Mulga Jack on Swans   52
Sacred Ridges above Diamond Creek   53
Amphibious   54
The Fifteen Hundred Year Vigil   55
The Slaughterhouse   57
A Stone Age Decadent   58

Acknowledgements

Many of these poems first appeared in the pages of the *Age* and the *Times Literary Supplement*. Others were printed in the *Bulletin, Clanjamfrie, Fiction Magazine, Island, Overland, Poetry Australia, Poetry Review, Meanjin, Scripsi, Southerly, True North/Down Under, 2 Plus 2*, and *Westerly*. Five sections of 'The Bits and Pieces' previously appeared in *The Emotions Are Not Skilled Workers*, published by Angus & Robertson. 'Mulga Jack on Swans' appeared in *Poems for Charles Causley*, published by Enitharmon Press, and 'Shadows' in *Lines from the Horizon*, published by the University of Newcastle. Several poems were broadcast by the ABC on 'The Poet's Tongue'. My thanks are due to all the editors in question.

'We wonder at ourselves like men betrayed.'
                                        WORDSWORTH

'he climbs the pineal gland as if it were a hill in the centre of the scene'   W. J. FOX

The Amorous Cannibal

Suppose I were to eat you
I should probably begin
with the fingers, the cheeks and the breasts
yet all of you would tempt me,
so powerfully spicy
as to discompose my choice.

While I gobbled you up
delicacy by tidbit
I should lay the little bones
ever so gently round my plate
and caress the bigger bones
like ivory talismans.

When I had quite devoured the edible you
(your tongue informing my voice-box)
I would wake in the groin of night
to feel, ever so slowly,
your plangent, ravishing ghost
munching my fingers and toes.

Here,
     with an awkward, delicate gesture
someone slides out his heart
and offers it on a spoon,
garnished with adjectives.

Mind

Along the face
of this baffling space-time construct
an illusory straight line goes ticking
fringed with words

in its long push
toward continuity.
Somewhere below
a chafed sea sobs

washing up the monsters
through tanklike dark
and a thousand fragmentary
shades of green.

The tank's, or football's, edge is flicked
by things of the world:
frond, blade, loop and wing,
tiny ecstasies of light

all laying claim
to a genuine status.
Odd packets of language rise
to acknowledge them

while displaced waters ripplingly break
on pain, loss or breakfast.
The plot is set on a small star's
fortunate planet

which we (so to speak)
entitle Here
trying to make
a map or globe of sense

with a child's
original box of paints.

The Figure in the Carpet

In recapturing awake
town-, land- and archiscapes
first dreamed
it may be ten years
ago, the self gapes
feeling of a sudden
plugged in to the true
presences, the deep Yes
whose warped hills,
roadweb, snuggling coombs,
halftimbered corner shops
and ferries have been
there since the first age
waiting for you.
But you twig
where the scenes came from
without a skerrick
of symbol-skill to crack
the code of their eloquence,
their bowel-stirring
music for the eye
of mental structures
and shimmering lagoons,
all gathered twice.

But you can't
quite make it
out from
the ground of
whatever has
supplanted burgundy
axminster and such;
neither is it
the tone of
trauma recovered
nor yet the
cousin of sleep,
but in the right
light it can
make your moist flesh
creep, the heart flop

freezing and the mind
race in neutral
without direction:
stabbing
you through and
back, like a drowned
language or
a bone voice
edge-on to you.

Recollection

for Gwen Harwood

Depends on how you scan the map
of memory, of chance and hap,
    which things
  come back to the one who sings
    and all, and all.

Suppose forgetting were the norm,
gullies not scarps defined the form
    until
  contours corroded each turfed hill
    and all, and all.

Struggling through scrapey, waist-high scrub
why recall anything? There's the rub
    yet these
  unbidden glow, light-gloried trees
    and all, and all.

The Landlord of Himself

'Every man is the builder of a temple, called his body'
                                            THOREAU

Soothing his pang by counting the tiles or slats
on the inner faces of a three-dimensional
figure with many a nook and ledge,
many a corner for weeping in

but which also here and there admits
some irregular mullioned panes
refracting bits of the world:
rows of hulking topiary, a glow

from summer's wide bronze acres
or just a girl getting into a car,
he trudges to and fro.

The panels are so glossy that they half reflect you.

In the long run he finds no room to take
he is eating his heart out,
the taste is oddly rich
but his shirtfront is tacky with blood like plum-jam
and his fat feet ache.

He paces past the spider's contribution,
a ghost of triangles and joinery
on whom dead forests close,
one lintel threatening to crack his skull.

The plot thickens, architraves brim with nightmare,
wailing a song about a barnyard fox
he plummets from floor to floor,
hands of carved angels letting him go by
down the builder's convenient smooth shaft.

Forgetting

Your philosophical moth
flutters against the glass
with hardly more than a shadow
of coarse doubt:

these nimble skipping images
are *they,* perhaps not even
reflected jags and fragments of
kaleidoscopic glassware

while anything tearshaped
runs terribly slowly
down the sheer
pane.

Abhorring a Vacuum

Pearlmother dawn. It is fairly true,
'The mind divests itself
of any belief in the mental'

but my slept fragments
fall back together into a shape
doing things with cutlery.

Outside the pane, frondage and dewdrop-cluster,
tiny birds in their twelve-tone clamour
recommend continuity;

no one I can see observes me,
who fade like blow steam.
Dozy-dim, I battle back as

the fiction of personality
nimbs me for a moment like
slant light through a tram door.

It passes. I am lived
by who knows what, the gene's blind way
of making another gene.

Whatever has been writing this down gets out
from behind the wheel and
walks away.

Smart as Paint, or Less

for Tim Bass

There are no names
for the colours that really matter,
our blueish tans of pity,
dark white
blood congealing on a purplish thigh,
and the future's matt blocks
of sandy deep-green.

There are no names
for the mingled pigments
that someone—let us call him P—
brushily applies to a blackbright thunderbolt
with my name on it.

The Shadow Minister

What are the black holes?
   They seem the incomprehensible
   wool of happiness
   knitted into a balaclava
   with staggered stitches.

Who mislaid them?
   I fear it might have been
   the shadow minister
   or gremlins in his entourage.

Who is he, then?
   Archduke of fractures,
   maelstrom, carcinogen
   and small betrayals.

And his credentials?
   Founded the vortex
   brown on the face of the waters;
   flavour of presence
   and overwhelming jonquils;
   a strong left arm.

What causes pain?
   The shadow minister
   whistling through woodwork.

Who knows the name of death?
   Only he has rehearsed
   its gorgeous nicknames
   and by decree can even call
   our frosty bones to heel.

   Listen closely,
   he is on tape
   in every dusty shadow.

Nub

Gro-ink. Kopita, kopita, ko-pi-ta . . . *konk*.
Were it not for the fact that I'd recognized the brickwork
of the previous station, coming out of my daze

I'd not have known. Out. I ran a whole block
at dream's peculiar slowness to catch my tram;
grey blocks of a city abstractly slid.

I came to the upside-down house,
its walls were folksified with Virginia creeper
and it looked as cheerful as a month of Sundays. I went in

being bent on finding the future, the logic of which
led me hotfoot up the stairwell to mother's bedroom
where, plumb on her chest of drawers, lay a single jewel-case

of tortoiseshell. Snap: open: there lay the future,
a perfectly polished chromium steel ball-bearing
round about five-eighths of an inch across.

I stared, closer and closer into its surface
so that the room, my face, were warped and bulgy
as through a fisheye lens, but rather fun.

In such a concentrating situation
what can you do but stare?—that's what I did
feeling immense waves of sadness in my legs

and chest all the while. I contracted the world's pain.
It was then, smaller at first than a bee's foot,
that something drew in closer over my shoulder,

an artefact as much as a consciousness,
jointed, threaded and cogged, full of spindles and heads,
a bronze-and-silver toy of elaborate construction

with a sharp weather-eye open. It knew the game.
It was the future. It gave away nothing at all
but a friendly tap on the shoulder. I shut the box.

Hoffmann Drunk

In that she spoke more
   about Don Giovanni
than of her role as revealed to me
    (masterly performance)
   and I could penetrate
     that foreign space
carefully recognizing its phantasmagoria,

she said her whole life was music
   and believed
  that the submarine secrets
   no verb can express
    singing can
     or was it
that now she could comprehend life through song?

While she was murmuring
   of the Don and her part
   it seemed that only now
the deeps of this masterpiece
    opened for me
and I could gaze lucidly into it
     recognizing
all the dislocating experiences
   of a weird world.

But who on earth was she truly and
    where on earth
  do we all go when we die?

Exit the Players

It is over. Self-mortifying Hamlet
gets up and puts his sword back on the shelf;
Laertes, unbloody, shuts his manila folder;

the Queen, who moved two separate amendments,
thinks Yorick played the fool offensively;
Horatio had the numbers this afternoon

and buffaloed through a policy decision
the Ghost would not have stomached;
Denmark is feeling just a wee bit whacked.

Ophelia's hair is dry; she didn't say much
but wouldn't mind slipping over to the pub
with Rosencrantz—why does he have to keep nattering

to the King? It's drinking time. Stiff Guildenstern
will not forgive Polonius and, to boot,
has lost his biro somewhere round the place.

It could have been worse. The dead all kept their tempers,
Gertrude cracked several, Hamlet one good joke
and they got through the whole agenda, perhaps because

Fortinbras is still on sabbatical leave.

Spirit and Act, or the Last Metro

The play rips on,
miraculous, gestus, bizarre
while the whole earth is footlights
but a weird message
hurdles the synapses
from heaven knows what cortical deeps.

The song's name is death,
actors driving home information
from the left lobe
or thespian bunker
where the invisible
decrees what kind of show goes on

and on and on.
Respiring through fleshy greasepaint
the actor is a standing cock
or brassy
barrel of jokes
but *the audience died long ago.*

Deep in dry earth
the director, the Hebrew,
ear to an airhole
sweats and thinks
while at ground level
Catherine's long thigh is getting stroked.

Squibs in the Nick of Time

1
Two heads
are better than one
said the professor
of pathology.

2
On the level
he went through
you know who
like a shot off a shovel.

3
I am cool;
you are too curt;
she is laid back;
he is as cunning as a leaning shithouse.

4
We try to work in the gap between
glücklicher Dichter and *obiter dicta,*
a narrow floodplain
overgrown with puns.

5
No-one speaks rudely, bar when pissed,
our verse is bourgeois-formalist.

6
Approving mystery
with all my heart
I practise disenchantment.

7
Do not call me a freak,
you scruffy dwarves,
I have designs
on the tattooed lady.

Redemptive Heights

Beyond Wits'-end a narrow strip of land,
Scarce quarter-acre, not desirable,
No room to build a residence,
Hardly to swing a cat
If that's the way your fancy runs:
Battles are fought here all the time.

A salt wind blowing back from Ragnarök
Rips through the pigface and the mean coprosma
Where neither gull nor gannet arrowing stoops
Over so much as water;
At margins such as this
The toughest samphire-picker would fall back.

Topsoil gives out. Cold hoofprints mill around.
These panoramic views are all of hell
Above which now you trudge, absurdly busy
With pegs and tape and ball of string,
Desperate at least to be at work
When the steep world stops.

Panoptics

It was another race
with a slender hold indeed
on the lintel of prehistory:
a graminivorous people,
gentle, bandy, greeneyed,
who left so little mark
on the tofts and crofts of time,
their adzes unremarkable,
shards pitifully few
and their passions blown away
like the long lavender wind.

They gave myth to the stars,
looked into that unbelievably
complex mosaic of
diamantine fires and
called the clusters by names:
there were The Kiss, The Groin,
The Phallus and The First Dream,
Climax and Dayspring,
The Hip, The Labia,
The Great Exploring Tongue
and Semen of the Gods.

It was another race,
gentle as rosewater,
libidinous like kittens,
who coined such poetry,
who got dusted away by process
to leave not even
their names on the starry chart
or in the sharp sand.
A new taxonomy
keeps all the stars in place
leaving our dreams alone
and the gone magicians.

The Good Spirit Bounces Back

1
Feeling is chocolate plus the
dry texture of a wolfskin
on which we sprawled by cosy gas
while mother unravelled
ivory knots of Chopin.

Memory: tales: a blonde, tribal
grandmother summoning her lad
with, When are you sailing over
to avenge the death of your brother?

Rancorous reason swans in whisky,
keels into work as dissipation,
castles a spectral king.

Bone bolts a scaffold-structure
to wedge these warring systems together.

2
At a dead loss,
dipping further this time
when you had thought him drowned,
gone deeper under surface
than pussy in the well,
vanished from your sight
or blotted out,
something
hardly perceptible
trembles again,
begins to rise through shadows
and comes once more to meet you,
arms outspread,
grinning like mad
and full of joy as a country fiddler.

Where did you come from this time?
you ask him now, not caring at all
what he is going to answer.

The Bits and Pieces

Swift's discovery, fundamental for art, is that there are no uninteresting objects in the world so long as there exists an artist to stare at everything with the incomprehension of a nincompoop.
                        ANDREI SINYAVSKY

Artichoke

Children's drawings of trees
approximate you
with your striated trunk
and dense neat head
inflorescent
of purplish scales or flakes
for our buttery meals,
seeming a sort of edible pinecone,
a green knight's club
or else an absolute rose.

Banana

The lordly nutritious banana
is peeping over the crunkled rim
of some old pottery bowl or other
up on a sideboard, at vantage;
look how his nose is black,
jetblack as lovebites that blotch
the yellow hide of his throat,
tropical patriarch
knowing perhaps
the new moon four days off.

City

When you enter some part of a city
that is completely strange
what excitement leaps inside you,
what delight,
milkbars and palms whirling like paradise,

each trim corner the colours of promise,
verandahs dancing,
stucco ready to please the eye.
I am good fortune
I am your future
I am the names of forgotten men,
it seems to say.

Dodgems

Metallic beetles
hum round the polished track
in quest of fresh crashes.

Emus

It is
particularly
the particular way
they come
stepping
warily
down the path
in dark
wrinkled
stockings
and shabby
mini fur coats,
their weaving
Donald Duck
heads
ready
to dip
and snatch
your ice cream
that appeals;
that
and the way
they browse dumbly brown
in cattle-paddocks.

Foot

Scalloped with toes,
flat as a hat
and pink as any nudity,
these ridges
and these shelly plates
will superscribe its crudity.

Fresh out of wool
as a hermit crab
it creeps across the carpet pile;
a mirror-sister
marches past
then falls behind it by a mile.

The range
of possible expressions
is limited to flex and kick.
An ankle canters
on this mare
whose littlest buried bones go click.

Galvo

Those large gaunt shearing-sheds
on the long tilt of
            halfshaven hills
are built of it:
they are pastoral cathedrals
of dull, rippled grey
or steely new ships of the soil
breasting against another winter wind
that fills the pines with whirling blackness.
The rustbrown battered stamping-houses,
dark ovens of history
            in the summer mountains
were made of it
when trolleys ran with quartz and gold.
Loosened sheets bang away
            in yellow Januaries;

    dark rotted pieces
            lie in the creek's edge.
In the backyard
a creepergrown green dunny;
at the township's edge
threatening humpies of the defeated ones.
Galvanized iron
            our modern thatch
for a long time you wore blue slogans
for Doctor Morse's Indian Root Pills
near this or that sequestered cowyard.

Hair

Heaping, coruscating, a waterfall
light enough for the breeze to leaven,
this all too human foliage
to fill a lover's hand, a coffin
or the red eye of St Paul.

Ink

With scratches,
splutterings
and with slow
determined blots
your thick fingers
learn to become learned
in it:
how to master
(Master Pupil)
the tyrannous
steel nib
into a painful
parade of copperplate
with no crossings-out,
into a decent hand,
the linear
life of the mind.

Jasmine

A presence—how can you name a smell?—
heavy, hazed, like peaches and honey perhaps,
spreading over the lawns and lanes;

look to the flowers, dense on their dark-green
pointed leaves and upcurving tendrils.
They are fivefold stars of white,

they have crooked points, they are
spokes of a series of wheels turning clockwise,
there is a touch of yellow inside the corolla.

Sweetness of jasmine,
it bends up from the rickety paling fence
by untended beans and caulies.

Can such floods
of scent
come from those frail starfish

bunched high on the tennis-court wire?
The gravel swoons,
the lobs drop back.

Kelp

Slowly it blackens
on the yellow shore;
a hardness thickens
more and more

in leaf, bulb, flange
and rubbery stem
along the fringe
or scalloped hem

of surf-surge. Time
turns all opaque,
including these

straps, grapes, trees,
fan, tress and rake:
gone their soft prime.

Lava

With glaring edges, lava crawled over the low
smoky plains, and stayed; slowly the Dreamtime
chipped it and crumbled it.
                                  Hardness decayed into loam
plumfull of bluegrey boulders or soft in the flank
of dawdling watercourses. Drifts of dark folk
came and went. Winds out of the west
hurry their squalls gusting over the sheep
new, canny folk know how to fatten and shear,
or change for cattle. Their hollow churches
bear the bleakness of that old hard basalt.
Violet, smudged evening: a vague dream of volcanoes.

Mars

Those permanent icecaps turn out to be built of water
and not of CO_2,
more yet perhaps tied up in permafrost
under a fine-grained red
speckled with dark blue
that clads this dream wilderness of the lost.

Cost, enormous. Our jigger landed then
and scooped up 5 c.c.
of sample from that painted desert—so
'We may be asking
the wrong questions completely'
but Tex's toy laboratory will know.

Krypton, xenon, hectares of rose rocks
minus 120F
but there is one thing to be boasted
of: it hasn't rained a drop

for aeons. Are the gods deaf
or is nothing in this wide universe wasted?

>   Nothing at all.
>   The planet rolls away
>   like an old, chipped compo ball,
>   Phobos and Deimos
>   asway around
>   his ruddy midriff.

Nectarines

From such rough barky trees
and crinkled crescent foliage they come
bearing the press of years
or boyhood's tang,
sharp, heady, themselves,
not this nor that
keeping their seasons happily enough
and turning one red cheek
toward the embrace of sun
or sharp white teeth.

Opener

is an astonishingly slim
metallic biped,
that is if you call
Captain Ahab a biped;
knees together while
his two teeth bite down hard
if you twiddle his ears round and round
and he skates round the edge
of a silvery rink
on his head,
in his frustration
biting right through that tin crust.

Plantain

Clumpy beside its crowded peers,
the mace-bearing lawn-invaders,
it has lots of ribbed leafage,
knows two hemispheres

and will not be dragged from the ground
except on a rain-soft day;
its fibrous pentagonal stems
are commonly found

just the thing for a sport
of conkers, bashing seedheads off in turn.
If any rooted survival
commands consent from thought

however it goes against the grain,
do not begrudge these haloes and fixity,
as green as any plantation,
ubiquitous like pain.

Quail

Quail in a beer-garden
brown behind netting
shy as the long drinks pass.

Roses

Oystergrey
thick rain off the vertical
on convolute roseforms large and small
a hundred kinds of blush
carrying fossil emotions
a thousand years or so
dangled from each wet petal.

Tenterhooks of rain gingerly depending
from that which is

Theatrical trophies
(melancholy queen)
bobbing on their spiny butchered towers
this densely implicated origami
for whose soft host of kindred colours
language has no words
                     (but surely there are words)
Curvacious, polygonal, rich, heartfilled cupcakes
when I bend down to the table
they slide away
into their rosy selves
                the one and the many.

Oystergrey
this hydra-headed showerfall
faintly clouds the score
of unbelievably resistant shapes and types
all of which mean rose
and are.

Skies

At Boolarra someone stitched the clouds
            to the edge of pillowing hills,
the swamp harrier, maybe, or the blue crane,
where dams like sightless eyes look up at stars or the sun
rolling above us blindly over and over again.

Telephone

Why does it drive me up the pole?
Perhaps my hankering for the whole
man, woman, child, in conversation
and not this echo of depletion
which murmurs through a diaphragm
following some branch-line tram
of thought in disembodiment
piles fuel up all round discontent.
Give me a presence, here to hand,
that I may view and understand
but not incapsulated voice.

Bad magic, this: a slice of noise
far from the circle of oneself.
(I wish it neither wit nor wealth . . .)
One's truest image stands close round
breathing upon Tom Tiddler's ground
and cannot be distilled for sound.

Underwear

On the whole they have been much neglected,
especially by writers, these necessary smallclothes;
figuring as Linen in the eighteenth century
and vanished completely of course in the next,
in our own time they seldom score a mention
except as trashy erotic accoutrements
to be 'impatiently ripped away'
in the more boring kinds of fiction.
Give them their due: they do yeoman service
to our crotchly comfort, our body temperature,
cradling of boobs and ease of movement.
Whoever has fidgeted through a bad day
in awkward underpants knows the difference
they make to our psyche, and learns to praise
a technology which has learned to create them
light, resilient, airy and snug.

Veils

I heard the suburbs talking to the stars
in a dialect I could not understand
full of dew, amber lamps, familial mortality
and Brennan's Wanderer fuming along the paths
of himself as the very first pale veils came down,
dovegrey, dovewhite, veins full of cottonwool
threading our crescents and the sometime creekbeds
like snow in the mind
                    tulle through a nocturne
                                    blurring on glass
till the sharp starchart vanished from my ken
completely.

Washing Machine

    Behind a glass pane
the wardrobe's wet personae
    play at vertigo.

X

You know that he is passing by
just on the barely other side
of marginal experience
by heartstop, a breath on your neck,
dim whiffs of garlic and old iron,
the ceiling of your mouth gone dry,
all fruits and leafage petrified.

Yams

As the crammed globe
rolls eastward on its axis
into the dark of history

carrying its usual load
of tyranny, art, gods, muddle
and excruciating pain,

it pleases me to think
of the humble yams
stacked in their market stall,

ugly, knobbled clubs
of mediocre taste
lacking even

that noble mythology
which has grown around the spuds
of Sir Walter Raleigh

and a million bags of chips.
The yam has no glamour at all;
it merely lives in the world.

Zephyr

Yellowish, crumpled, frail,
the leafboats lie
on water black as hair
under a marbling sky;
the secret wind moves through them
unseen but visible.

Puck Is Not Sure about Apollo

Being rather tanned and rumpled,
rusticated as a thorntree
I don't quite know what to make of
Mister Smoothie with his sunblades,
general air of curls and music
plus the occasional golden handshake.

    I do not always like this;
I lean to whatever is tousled,
windy, brownish, water-kissed,
lit by irregular streaks of glinting
and, above all, out of place.
Allow limbs, thunder, peeling bark,
fish that splash out of rank in rivers,
crescendi (or are they crescendoes?),
heartleap, pulsebeat, jape and joke,
antics of self-dismantlement.

Choric odes don't seem to be
my especial cup of tea,
who skip and whirl across the map
like a Tam o' Shanter sort of chap . . .
all that hellenic sunshine, ah,
it almost goes too far
or maybe not,
              OK?

The Fall of the West

Jig, barbarian, over these
machicolated ruins;
question your soothsayers what they mean,
these bricky brac geometries
and bushland scene.

You are standing, marauder, on a college—
a kind of, sort of barracks
plus public house where the gone pale folk
laid claim to worship knowledge.
See, where that mouse

scurries through rubble was inscription-barn
in which they stacked their lore.
Why did they vamoose and let the bush take over?
Why leave this extended pinkish cairn
without dish or door?

Ah, that would be telling. The gods have not replied
but let us say this wing
was named for Boog who devised the manila folder . . .
   Just looking at such ruinous things
   I feel abruptly older . . .

This middle part for Kogo, inventor of scissors,
that mound beneath low trees
for Pipp who first designed the bike and the bra,
and so on; its deeply graven logo
is too arcane by far.

Right, I was coming to that pointy-crested minaret,
no doubt proconsul or priest
bawled from its holes his awesome guttural words . . .
   Look, in the shrubbery, that odd beast.
   And what are these bright green birds?

Eating the Future (I)

Can I feel order in the whole shebang?
Pain I experience, and blood, and phlegm,

hardening of iliac and hamstring;
but as for order—I write verse.

Something is larger, like role of light at noon
enabling turquoise, lime and cinnabar,

we feel its brushy mayhap and we sing.
I stroll abroad as Puck, wagging or winking.

The city that I thread through is a flower,
the clouds are escapades of cottonwool

which give aesthetic cuddles but we hurt,
knocked rotten by the blues of random power.

Without god the upshot turns out worse
but with his aid it cannot make much sense.

What did Masaccio think of it all
before his red paint dried up very young?

Trees blow. We all feel shallow. This bay glints.
The speed of light is constant everywhere.

Eating the Future (II)

I looked down in amazement,
slumped on my festive seat;
I had eaten the future entirely,
sauces, garlic and meat.

I stared at the ruined table
in paradoxical joy,
having swallowed the whole fat roast
and chucked the bones away.

There were still grand wines in the pantry
and vintages of the past
but the voice of the party crooned away,
'Nothing is going to last'.

There were bones all over the floor now,
yet Eros stood decked in flowers.
The tablecloth map of potential
had shrunken from decades to hours.

Though lights go off at this party
our jazz is blaring loud.
I seem to have eaten the future,
hilarious, unbowed.

Words

go drifting
hither or yon,
helter-skelter or slow,

word magicians
or old-style dealers
moulding the stuff like dough.

Verbs look back at us,
even create us,
backbone and furbelow.

Do they know
all that they claim to know?
Probably, no.

In the beginning it was, however,
like a darkblue unfolding
O.

That Radical Politics Is Impossible

You'll grow up,
you'll settle down,
they'll boil you down
into the common toffee,
knowing this:

someone has to scrape up
the fifty-one per cent
of requisite agreement;
somebody has to make
the whole caboodle work.

There will be no point in asking,
'What was my gift?
Where is my true note?'
We're happy as larry to pay you
not to find out.

Don't rock the boat
or you'll crack the bloody duco.
Don't bite the managers:
they were the boys who handed out
these oars and rowlocks

which you may choose to ignore.
So belt up
or put up or shut up because
to him which hath some gravy
shall gravy be given forever.

Sand in My Shoes

Dreams of the dour and winsome,
a world for ransom
and a stammering impresario
by the piano.

Ah, the American magazines,
the flagon red after light toil
and summer's fading masterpieces
cool as coconut oil.

Practitioners of Silence

So much that happens happens in
the gaps between, the spaces,
so much meaning cries aloud
from what you didn't say

or else he didn't
or she was looking away
thinking the quarry on the hill's blue flank
had a kind of truth, too.

Take those exchanges: 'G'day.'
                               'G'day.'
'I been thinking.'
               'Yeah?'
                          'How're the sheep?'
'Not bad. Been dry, though.'
                          'Need a few more?'
'Maybe. How's the truck running?'

and so on, at the gravel margin,
vibrations hanging in the air
like a black angel with a fiery sword
or a mortgage spouting blood.

Between the stanzas, under the words,
whiteness like chalky bones
as orange clouds float over, one by one,
rhetorically immortal.

We'll Build a Stairway to Paradise

There is curious distinction
in that ladder
running up into a tree.

One looks in vain
for a plump couple
fornicating in the forks

or for some leafy staircase
feathering right on up
to lucid puffball heavens.

It appears
a wooden conception
of ascending;

absurdity
props up
its painted slant shafts

but one could fancy them
the bonded stilts
of a prestidigitator god.

Dark yellow
parcrumpled leaves
glide glumly groundward.

The Home Conveyancing Kit

Michael and Trish have bet their bottom dollar
on getting the house paid off, it is so sweet
they just fell in love with it from the start

so bought it, or began to, from Mario
still paying it off himself
who had a malignant growth removed before Xmas.

He it was planted the vegie garden
after handing over the deposit to Hugh and Min
who went overseas to try and save their marriage:

it failed, of course. The paint job and the sleepout
were the work nine years ago of dotty Trevor Platt
out of whose estate the balance was paid.

He had snapped up the then-quite-newish cottage
from a wolfish pig of a landscape painter
who mainly cared for what was under skirts

and collected it—till he suffered an amputation.
The place was originally sold by a dairy farmer
who never made it back from Armentières

as a building block within cooee of the station
before the city sprawled and gobbled it up.
Under the lemon are the bones of somebody's dog.

Patient Portent

Eleven Jerseys on the brow of a rise
stand gazing into luminous west
as it might be waiting for godwit

since their horned heads are all fixed facing
those blurry smears of westward nimbus
toned like a galah's wing.

Eleven Jerseys loom against the sky
on their melodramatic knoll
but what will be the raiment of . . . her, him

or it when godwit comes,
blatant as any flying object
and radiant like flushes of puppy love?

If any source came seeping up from the creek sands
to alleviate world-pain, would it speak
with tongues that got through to us
or might it say something grassy to those Jerseys?

A steep mound, with silhouetted horns.

River

Springs of course
high up,
skittering out from innumerable
riffled ridgeloads of virgin timber
yet unspotted by woodchippers
on the new make,
it skirts the ruined vineyards
which met phylloxera
and came a cropper,
sandstone, tan clay, Murray Greys browsing,
winds past medical ranches
to scrubby suburbs of mudbrick, Volvos
and odd literati
hoarding their dirt roads,
soon slowly weaves
into uppermiddle Christendom,
motor mowers' ratatat
and vigorous E. coli,
tram, bus and steel bridge salute it,
Italians on tennis courts,
golf, kayaks,
the halo of truck tyres,
a sluggish umber, it quits
the palazzi of education and reform
for factories in rusty reefs
shored up by protection,
fanfares tootling through the city
and real oriental ships
wearing its bottom on top
as the northern joke says,
trying not to be
poetical,
this is the Yarra, Smetana,
shrunk from the Yarra-Yarra
by linguistic diminution
and coiling at last
andante
into the sea
or a gunmetal bay
in a rippled fashion of dying.

Kia Ora

I always like to think of it as a hipped,
shambling, roundshouldered house, the boards
pale-weathery but many rooms full of nice
unpretentious comforts: the seats for sitting in;
record-player scratching away at Mozart
and polymorphous Fauré—estive sound;
iced water in the fridge; bright scatter rugs
on a stained floor; heaps of magazines.
Pearl and Rowan must have subscribed, you thought,
to everything from *Westerly* through the *New Yorker*
to the by-now-monthly *Women's Weekly* (for recipes).
They could damn near end up buried in newspaper,
they and their piebald mongrels, Tycho and Splat,
plus that reginal black cat, Full of Beans.
Nothing which entered that house ever left it,
you would have thought, though shattered logs
went up as winter smoke. It was shaped
like memory, ramshackle as a soul.
You came through the door, walls murmured Good Fun
flashing bold prints at your eye, for example
van Gogh, Lichtenstein, *Casablanca*, Fred McCubbin,
you know the scene . . .
                        Tanya thought it
the kind of place where Mind took off his hat
in quest of meditation, lemons, cake,
while Eye unfilmed her lens. Whoopee,
this is the shot: I reckon I never saw
quite such a realm of timber, glass and wool
nor ever such springy buffalo grass outside;
you felt you were in the twenties
you felt the world had never left the twenties,
harmonious as apple compote. Past
a ragged plumbago hedge all cars were filtered
out of our view. Hubbub was internal,
bare feet mandatory. Those kids would bawl,
            There's a track
            Winding back
                To an old-fashioned shack . . .
grinning like drains or chucking unripe plums

before they picked up teams for tippenny-run
against the wall of the shed:
                         I BAGS IN FIRST.
I always like to think of them all being
familiar and transient as newspaper;
dunked in tradition like pittosporum.

When rock cakes are brought out on a tin tray
round about four, between the snappy fuschias,
halting our game, I want to say,
'Listen, the world ain't like that any more',
but haven't the heart,
                besides, I like the rock cakes
and a large enamel mug of sugared tea.
Thank them all. Go. Maybe the liquor of history
is a dream into which we're trying to awake
like cicadas. Esther calls out, 'Bye-bye.'
Bye-bye, then.

Shadows

Twice in Manhattan I spoke with my double, Death.
He and I wore the same herringbone overcoats,
were sallow, dark, same sort of age and build,
but not identical: Death is not naive
though always banal (a word I treat as an iamb).
Our first confrontation was in a Doubleday bookshop;
he looked straight up and said, 'You can always trust Fiedler',
then went on raving about the Boston Pops
till I snuck off. Chill, smog, a fortnight later
I bought a thin cigar and the New York Times,
stopped in the shop door, and there he strolled up again.
Same tweed coats: same curious quantum leap.
'It's hard,' he said, 'to get music for the triangle.'
'It is,' I replied wittily, never saw him again,
missed the flash, the sulphurous smoke, the golden
trumpets in mad crescendo; not a squeak or a peep,
only the double and triple helix of fog
between Park Avenue's perfectly stupid glass towers
unwound, knotted, blent its merino skeins.
I wouldn't want to see him again for quids.
And my father spoke to his döppelganger once,
coming out of the jungle in '41
to a dead rubber planter's haunted house.

Aunt

While your droop skull lolls sidewise
on a twiggy tree of bones
you manage a few hauled phrases,

Violante,
longdoting mentor,
riddler of kidhood, rug-hooker,

knitter, cook, hatter,
camouflage-net knotter and impresario
of cardgames galore, reciting

'In Parliament House when a motion is made
Each member stands up and uncovers his head
But in this little house it is just the rev*arse* . . .'

etcetera, gold teeth wide agrin,
benevolent also to your fruitful hens
but sharp as a tack about neighbours,

notably on the catholic,
the palpable reffo
or the drinker of bottled beer.

Your crony, Madame MacPherson
had been at the Siege of Paris
eating rat and dog; our relatives

remembered even the Mutiny,
there were links with Nicol Sain,
brown snaps of Cawnpore and Lucknow.

We were tribal Scots,
the Zulus of Europe, gone southern,
your jokes were all fair dinkum:

you had, I long remember,
an imaginary dead fiance,
a digger called Inkpot.

Then, you could do everything except swim,
and yarn about everything else,
so I twigged most tricks from you

but now your head lolls bedwise,
a heavy fruit, bride of death,
and your skin doesn't fit.

Under the Roof of the World

It was I who pounced on the back of the burly werewolf,
struggled and cried, pinned it at last to the ground,

black fur flattened in backyard, held it down there
unremitting, tightened my grip, choked out its puff

and broke it in pieces like bread. Those bits lay there,
bearlike pelt, claws splayed in pearly-grey dust.

Up I got: thank the powers: it was clean over,
time for a scotch so I turned on a heel and went.

But no, godsbreath, NO, I found they'd forgotten
to burn those scattered hunks of the dead beast.

More whisky called for: come night, sure as day
the pieces would all join up, spring back together

as filings leap to a magnet. My weird would rise again
following nose toward me in even the pitchiest blackness

in need of a bait, of a death, of me-flesh
waiting where tears of stars drip down through a ceiling

onto the bones' sleep, slow drip of the pancreas,
tendons and string gone lax but skull containing

germs of a tale where by an unpainted cottage or house
it was I who crashed on the back of the writhing werewolf,

his throat soft under my fingers . . .

Gaspard de la Nuit

Are these all many or is it one?
enquiries that revolve,
move the feet and thump the pillow
round about 2 a.m.
with pale blue moonlight splashing on the carpet
and a mopoke moaning somewhere
among the pokerwork hilltop trees,
odd cars bent homeward
distantly.
One or a medley,
night, nuclei.

Our galaxy has no point at all
nor do the others
gleaming down
on our marriage, bills, piano,
on Parliament House,
T-shirts and ammoniac nappies
lodged in their chink in space-time
for some directionless reason
soon to be rubbed out like a blackboard lesson.

Sleep now,
dreams are much easier
and cheap as chicken broth.

Still

Pink top
magenta jeans
you rake the crop of aromatic
newcut grass to mounds
hedged in by hedges
vividly composed
as in a French film
of shimmering bourgeois
country highjinks.
Five years together,
birdcalling together
our harvest home:
beyond that squareclipped cypress
the dark green sea roars.

The Mower against Ocean

On my side of the highway grow
the headstrong grasses, knot by knot
and row on row.
I roar them down with my
old Briggs & Stratton juggernaut.
I shave back short
kikuyu, fescue, bent and oats
but at my side, aloof, distrait,
the multicoloured ocean stirs
illuminated by indifference
through turquoise, indigo and jade;
ocean can not be mown,
would snap me up as soon
                      as Bob's-your-uncle,
gently growls and rumbles
at behest of a moon,
eludes iron's edge.
Let flesh be grass, that's how it is,
let dandelion and black-eyed Susan fall,
sea is a cold salt blood.

Mulga Jack on Swans

for Charles Causley

He saw a white swan first when he was thirty.
    The graceful chook
moved on a pond in a provincial city
   to which it came, I'm sure,
     out of a book,
     from literature.

How clean it was, how primly Persil-white,
   what a good colour
    for swans to be,
replacing their habitual shade of night:
O brave new world, that made such birds,
    said he.

Sacred Ridges above Diamond Creek

for Les Murray and for my pocket tape recorder

I want to make some kind of gesture of alien response,
response no longer alien, response finding its feet,
salute with my feet and my hands and heart to the totem beasts,
by whose names this district was once patched and pieced like a tartan,
to echidna, formed like a child's drawing of an explosion,
to dingo, long driven away by Mr Fox in his red hunting coat,
to platypus, the shy, the watery secretive,
to magpie, sweetest yelp of the morning,
to sober-suited shrike-thrush, epitome of musical variation,
dour tortoise like a stone
and vibrant kingfisher smart as paint.
These are thinskinned sandstone hill-rows
remaining shaggy with yellow box and stringybark
despite their undercoat of heedless houses,
and the gutters still gush to tanglethroated creeks
so I pay my spilt tribute to all those neglected totems,
to clever-tailed possum with his unspeakable voice,
to willy wagtail in flight, a whole symbolic system,
to kangaroo, whose name at least we dared to retain here,
to square koala, cosily hunched no longer in manna gum forks
  along Main Road,
to black bream and snake
(the latter perhaps renamed after an Indian quadruped),
to kookaburra, raucous herald of fire's daily return,
to the wild cat, whom I do not know,
to the 'furred and curious wombat',
to you all, primal kin of the region I choose to live in;
each genius loci now displaced by incomprehensible names and grids
though the cunning birds, secularized, remain,
I yield you primacy now,
you and those adamic makers who prowled around these
  scrabble-shingled hills before me,
whom we may choose to ignore at our own cost,
hail and farewell, but then, after all, hail.
Yours is the first magic.
Yours are the names of things.
I cannot divest myself of this curious tongue
but I lay down my arms.

Amphibious

Shall I say I remember, sort of,
the way a body remembers things,
two or three stick-in-the-mud steps
at an edge, then stooping in,
pushing off across that toffee-toned
slight swirling of leisurely water,
its barely discernible taste of earth
(which is part of being primaeval),
to and fro breaststroke, touching down
now onto softness, now on the smooth
bumpiness of water-fingered pebbles,
and once reaching up into
the lighter overlying medium
to pluck out of green shade a spray
of ever so gracefully descending
rivergum foliage?
                Yes, I shall say it
with all the fervour of hands and feet
quite unaccustomed to abstraction,
feeling for those curved leaves now.

The Fifteen Hundred Year Vigil

Still they maintain their monotonous observation:

five mathematicians climb every night
up to the summit of their tower
alert for what goes rolling overhead;

one stares rigidly into the zenith
and the other four
at silent compass-points nightlong

so that nothing of what goes on
in the sky's four corners
can 'scape their diligent diligent observation'.

       I see them now
in burgundy silk robes
covered with fabulous starlight.

From unbelievable warp or deep
the universe looks down
and spies on them

or lays a black frost
on miles of cabbage fields.
They set great value on glass:

so intricately bottled up
as nearly to be invisible
the sand
       the soda
              the silicate

all now bitterly bonded
though capable of splintering, and quite
as hard as any old heart.

They can see clean through it
or it glares a face right back;
at orange sundown my Hurstbridge train

is a clanking corridor of mirrors
burning with illusion
as dark trees race through glassed heads.

It was when the knowledge-mirror broke
that there lodged a jagged fragment
in everybody's breast

which cuts and hurts like hell.
When I peer in the shaving glass
my thick lather darkens with blood.

One morning, doing business in a city office,
I saw the suicides of the last few years
rain gently past the window.

The Slaughterhouse

for Katherine and Jim

It is shaped like a toy version of a man's head,
has white hair, grey face and a large red hat.
Open it: it is eight feet by five
under its gable roof,
its brain cavity carefully lined with dirty flywire.
Six meathooks hang from galvanized iron bars.
How neat and tidy it is, if dusty.
It has two shelves, the lower large, the upper small,
two saws, one small and one enormous.
It contains a stout, well-hacked chopping-block
split down the middle, inside it
a tunnelling spider's web.
The inside is cream with a concrete floor.
The focus of all is Diabolo No. 1, a pedantic mincing machine
which can whirr like the Primum Mobile.
On the upper shelf a small dusty drink bottle
is lying on *The Yass Post*, October 8, 1980.
Outside of all this
beyond the louvres
ears tell us
cockatoos and galahs exist
and through the louvres
eyes take in
dry offwhite grass.

Open the face.
Step outside again.
Under the peak of the red hat
there sticks a small hunk of honeycomb.

A Stone Age Decadent

Uh.
    Uh.
        These errant stripes of sun
That feather in play across my legs design
Transient ochres, ripples which the Sea
Has lent the air. I seem to like it here:
My tribal brothers work back up the stream
For tucker; one small knot of womenfolk
Go gathering shellfish where gold sand meets rock
There in my middle distance. They compose quite well,
Sun-burnished nymphs and mothers with dry tongues.
Under these casuarinas on my slope
Of sandstone and soft needles I may hold
A laid-back peace, keeping my cavernous head
Well stocked with pictures.
                Mm . . . hmm. Let them retain
Their rules and moieties. I disturb no-one,
Neither affront those boring Bluetongue rules
Nor trespass on Echidna's blunt decree
Here where a sea-breeze lightly lifts my hair
Flavoured with faint salt. Tribes are wise enough,
Let them think me no-hoper if they want to,
It troubles me no more than bushflies do
While thoughts waft up this hill: from here old Sea
Is crinkle-turquoise—rumpling, ruffled white;
Gulls and swallows thread it.
        Uh.
              Sheer solitude,
Watching these topmost branches bar sky's wink
With their shockhead tresses. This is my secret
Adaptation of totemic ground,
Sequestered high, brown half-shade where I lounge
Sending my spirit out to meet the Sea.
Fly, fine colourless bird, on thinky wings,
The words we use are only the words we're given,
They do not like to hear me saying that,
Preferring old songs, with their boom-bam-boom:

> *Here the big wave runs upon the shore.*
> *Here the spray blows up and up like smoke.*
> *Daily the shellfish, daily.*

Young girls gather foodstuffs by the white sands.
At night the seagull has stopped crying
Daily the shellfish, daily.

Totem and law, laws and restrictive totems,
Banal, sublime, bestial, that's how
My fellow tribespeople make out the world.
Practical sure enough—the food comes in—
But bone between the ears; at their sheer best,
Witness down there, say, brownbeautiful yet dumb.
Motes dance in light-slant just above my shin
And the glow-filled Sea flakes off her thousand colours
In tides of mystery.
               Listen.
                       Gull and currawong
Sound their antiphon. How fast the shadows
Lengthen on sand, coarsen the hill-textures.
There is pleasure in it all if you sit still.

I do not think they like me very much,
Not even Moama with her small round breasts,
Scrub of light curls, pool eyes, fastmoving limbs
And buttocks I could cup in these two hands
And then . . . Oh-oh! It's very much too pleasant,
That's to say painful, this line of thought:
Body responds. There! She stoops at the rocks.
I see only a single cloud today,
Thin, flattish, grey-white, drifting above the horizon,
All else says blue meets blue. And I relax
On springy casuarina needles here, my den
With a view. A spinebill's vivid uniform
Flashes to flowers a little down my slope.
I flex, reflect, withdraw. Ah, me. We all
Must learn in a line of days to wither up
And die—or else die first. Just like the scallop,
Mussel, periwinkle, any living thing.
Ahi! Know something now. Am I a fly?

Here the spray blows up and up like smoke.
Daily the shellfish, daily.

Peer closely at these jointed leaves or branchlets,
Green fingers of slender skeletal hands
Knobbed with small tan knuckles. Just to stare
With care at this or that makes world seem good,
Be it spiky conelets or multiply-scored bark:
I like it here. Those women on the sands
Make up a dance that fits a larger dance,
The bay, the hills contribute to my joy
As I do nothing. Ha. Yes. That's my game,
My hunt for needful store of images.
Lovely, yes, but what substance underlies all?
What might all change mean? Are we like shellfish
To be shucked and eaten? Why does the great sun set?
I wonder how we tagged these words to all.
Life is more than animal grease and ochre.
I well might fall asleep . . .